MW01592823

Birds of Prey

and Other Male Species

Male Chauvinist Pigeon

Dodos, Birds of Prey

and Other Male Species

An Uncommon Field Guide to Male-Watching

Mary Taylor Gray

Fulcrum Publishing
Golden, Colorado

For their humor and ideas which helped make this field guide a definitive reference for the art and science of male-watching, thanks to:
Carol Gifford; Ginny Hilton; Leigh and Randy Johnson;
Laurie Paternoster; Judy Sheppard; Sonja Short; Jon Talton;
Patty and Tate Taylor; Sally Taylor; and Rick Young.

… And others who unwittingly contributed great material.

Copyright © 1994 Mary Taylor Gray

Illustrations by Christopher A. Boyer
Book design by Patty Maher

Library of Congress Cataloging-in-Publication Data
Gray, Mary Taylor
    Dodos, birds of prey, and other male species : an uncommon field guide to male-watching / Mary Taylor Gray.
        p.   cm.
    ISBN 1-55591-182-X
    1. Men—Humor.   2. Bird watching—Humor.   3. Birds—Humor.
I. Title.
PN6231.M45G72   1994   94–17570
818'.5402—dc20   CIP

Printed in the United States of America
0 9 8 7 6 5 4 3 2 1

Fulcrum Publishing
350 Indiana Street, Suite 350
Golden, Colorado 80401-5093

Species List

Game Birds

Songbirds

Universal Species

How to Watch Males

This field guide is meant as a primer on the various species of males resident in North America. Users will find descriptions of plumage, habitat and lifestyle to help them in identifying the often confusing males that occupy the world around us.

Male-watchers should seek out each species in its habitat, then observe closely. Certain calls and actions on the part of the male-watcher may elicit identifying behaviors. The wearing of provocative clothing, raising the hood of the car or acting "dumb" may all bring out the true nature of a male. It may be necessary for the male-watcher to engage in and pretend to enjoy certain activities, such as watching football on television, riding motorcycles and listening to the male talk about himself.

Necessary Equipment:

Feminine wiles
Feigned lack of interest
Patience and persistence
Stiletto heels or athletic footwear (depending on species being sought)
Butterfly net and steel leghold trap (optional)

The Art and Science of Male-Watching

Male-watching is an art as old as the oldest profession (certainly some of the skills employed in one are used in the other). Undoubtedly it began in the days when pre-*Homo sapiens* lived in small bands on the savannah.

Imagine a hot, dry day on the African plain. A group of cooperating females is sitting around getting the job of living done—finding food, feeding infants, gossiping … and watching with patient resignation as the males, their newly acquired upright posture freeing their hands for uses never dreamed of before, grab clubs and hit each other over the head. Certainly those semi-simian females must have looked at each other, then given a shrug which even today means the same thing, "Men, who can understand them?"

Male Species Diversity

The development of identifiably different species of males is a classic model of Darwinian evolution. In the beginning there was only one—the Typical Male, *Y-chromosomal americanus.* But as a changing environment exerted pressures on him—television, professional sports, the internal combustion engine, women in the workplace—he evolved and adapted into the diversity of types occupying the many niches of modern American life.

Characteristics of Males

All males share certain general characteristics, both physical and behavioral. The most interesting is the location of the brain between the legs. This results in a preponderance of behaviors which make no sense to male-watchers, or the females of the species. It also results in male behavior often being determined by physical rather than common sense considerations. Male-watchers should keep this in mind whenever dealing with males.

The second overriding trait is the ego. Obsession with ego often drives males to act in nonsensical ways that on the surface seem nonadaptive. This may take the form of enacting a highly dangerous activity on a dare, or spending exorbitant amounts of money on a two-seater sports car when the male has a family with six children.

The third characteristic is the tendency for flight. Overt interest by male-watchers can startle these timid creatures. Male-watchers are well advised to maintain a low profile and not express enthusiasm at the presence of any male.

Note: This is a guide to watching *males. The capture and collection of specimens is an art and science of its own, and we could not begin to cover it here.*

To the males in my life—past, present, future.
They keep things interesting

Birds
of
Prey

Noncommittal Nuthatch

Noncommittal Nuthatch

Latin Name: *Commitus phobias*

Call: Look at the time. Gotta run!

Plumage: Usually wearing running shoes and carrying suitcase

Habitat: Halfway out the door

Behavior Display: After spending period of time with one female, acquires caged-rat look. May disappear when female starts to respond to courtship display. Becomes absent-minded about dates, especially when meeting family of female. May move into nest with female but keeps own bank account, phone number and apartment. This temperamental species is easily startled by any mention of nestlings. Flight response elicited by the words "commitment" and "marriage."

4

Heron Chest

Heron Chest

Latin Name: *Machismo testosteronus*

Call: Bay-bee!

Plumage: Tight-fitting jeans to accentuate crotch, muscle shirts worn over puffed-out chest or shirt unbuttoned to navel to reveal chest feathers

5

Habitat: Ubiquitous

Behavior Display: Struts and swaggers, especially in presence of females. This species is certain all females are fascinated by him and receptive to his attentions. He is impervious to clues that they aren't. Prone to challenge other males to contests—who can belch loudest, spit farthest, chug beer fastest, eat the most brats. Display stops just short of chest-beating. Never eats quiche.

Valentino's Warbler

Valentino's Warbler

Latin Name: *Latinus loveri*

Call: You look mahvelous!

Plumage: Tight black pants, breast feathers unbuttoned to navel, head feathers slicked back

Habitat: Bedroom is most common habitat, but also favors chaise lounges in boudoirs or other areas suitable for reclining.

Behavior Display: Extremely attentive to female. Ritualized courtship includes smoldering glance followed by excessive flattery and hand-kissing. Highly polygamous. Frequent mating. This species is primarily nocturnal but given to occasional nooners and afternoon activity. Swoon-inducing display consists of muttered terms of endearment combined with use of "bedroom eyes."

8

Bikerhawk

Bikerhawk

Latin Name: *Harley davidsonus*

Call: Hey mama, wanna ride my hawg?

Plumage: Black leather jacket or jean jacket with sleeves cut off. Vestigial wings reduced to Harley Davidson logo on back of jacket. Other identifying marks include motorcycle boots, chain from belt to wallet, long stringy hair, sunglasses, tattoos and pot belly.

9

Habitat: Highways and blacktops across the United States. Migrates to Sturgis, South Dakota, for a few weeks each summer.

Behavior Display: Usually seen traveling in flocks astride motorcycles with females perched up behind. Roost sites of bikerhawks are easily identified by the cluster of motorcycles parked out front. Display of revving motorcycle engine functions similarly to bicep-flexing and money-flashing in other species.

Midlife Chrysler

Midlife Chrysler

Latin Name: *Afraida ageing*

Call: I need to find myself.

Plumage: Head plumage combed from just above one ear across top of head to hide bald pate. Shirt unbuttoned to expose chest and gold chains. Mimics plumage of males who are just beginning to shave.

Habitat: Discos, health clubs, hot tubs and sports car dealerships

Behavior Display: Passage of young female elicits sucking in of gut, puffing out of chest. Drives Porsche or other sports car. Abandons long-time mate for the company of very young, blonde chicks of any species. Mimics behavior of juveniles. May take up skiing, tennis or sky diving, despite physical condition.

Horned Stud

Horned Stud

Latin Name: *Hormonus anytimus*

Call: She wants me.

Plumage: Tight blue jeans with pair of socks stuffed down the front; condoms in a full assortment of colors

Habitat: Trendy nightclubs, health clubs

Behavior Display: Strutting walk. Obsessed with large appendages (real or imagined). Body temperature always cool. Excessive hormone production. Refers to female with the diminutive "babe." Sight of female of any age, species or appearance stimulates pursuit. Obsessed with mating, especially when shallow and meaningless. This species often only has a first name.

14

Greater Spender

Greater Spender

Latin Name: *Money no-objectus*

Call: It's on me.

Plumage: Flashy suits, diamond tiepin, gold cuff links, gold chains around neck, cellular phone

Habitat: Expensive restaurants, nightclubs, limousines

Behavior Display: Strutting display elicited by presence of attractive females and other successful males. Flashes rolls of money, tips big, makes elaborate show of "picking up the tab." Typically accompanied by young chicks with very flashy, expensive plumage and exaggerated breast feathers. Mate usually swathed in skins of fur-bearing animals.

Disco Kinglet

Disco Kinglet

Latin Name: *Saturdaynite travoltas*

Call: Stayin' alive!

Plumage: Shirt unbuttoned to navel, fake gold chains, tight polyester pants, shoes with 2-inch heels

Habitat: Discos and nightclubs, mainly on Saturday nights

17

Behavior Display: Though during the week this predatory species blends into the general populace, it emerges in full plumage on Saturdays. He is nocturnal and afflicted with Saturday night fever. He prowls his nightclub hunting grounds, evaluating females. Upon identifying a prey animal, approaches casually, hoping to lure female onto dance floor. Hip-swivel dance display involves exaggerated sexually oriented gyrations; ends with finger pointing toward ceiling. Much emphasis on external tokens of masculinity.

Note: Many experts thought this species had become extinct at the end of the 1970s, but numerous remnant populations exist.

Woodypecker

Woodypecker

Latin Name: *Beaver overeager*

Call: Come on, pleease!

Plumage: Always tries to shed plumage

Habitat: Parked automobiles, movie theaters, darkened living rooms, booths in dark corners of nightclubs

Behavior Display: After buying dinner, this species enacts the "coming on" display which may involve pretense of stretching followed by putting arm around female; putting hand on female's knee, then moving it slowly upward; or overt grab-and-grope sequence. Constant attempts at mating include whining, pleading and bargaining with female for favors.

20

Bluejerk

Bluejerk

Latin Name: *Whatta creepus*

Call: Never when he says he will.

Plumage: May appear in various guises of plumage, though all eventually moult to camouflage. This species is identified more by its behavior toward the female than by appearance.

21

Habitat: Found in all habitats, except where he's promised to be.

Behavior Display: Highly attentive when courting female; disappears promptly after mating. Returns infrequently to indulge in cajoling courtship, then disappears again. Afflicted by memory loss in forgetting dates and the female's phone number. Also unable to tell time so is often late for dates. After pledges of undying affection, this species is frequently found in the company of other females.

Male Chauvinist Pigeon

Male Chauvinist Pigeon

Latin Name: *Sexist dinosaurus*

Call: Fetch me another beer, wench!

Plumage: Takes on many guises

Habitat: Found everywhere, including the Supreme Court

Behavior Display: Likes to tell dumb blonde jokes; pinches tail plumage of females; never notices that females are not flattered by his overbearing attention. Expects female coworkers to make coffee and answer his phone. Possesses greater knowledge about all topics than any female in his presence. Feels compelled to tell females how to do everything. Refers to women as "girls," even when they're sixty years old. Known to remark with amazement, "She's smart and pretty, too!"

Married Magpie

Married Magpie

Latin Name: *Cheater infidelitus*

Call: My wife doesn't understand me.

Plumage: Plumage may vary widely but identifying field mark is untanned band of skin around third finger of left hand.

Habitat: Bars, offices, cheap hotel rooms

Behavior Display: Given to obsessive behavior over females other than mate. Frequent furtive rendezvous in motels. Unable to be called at home. Despite frequent promises to migrate, this species always returns to its mate. Endless excuses for why he cannot leave the nest—nestlings, health of mate, shortage of worms. Though he claims complete lack of mating activity with mate, she just hatched her sixth chick.

Carrion Eaters

Black-Capped Cheapadate

Black-Capped Cheapadate

Latin Name: *Gotit neverspendit*

Call: Let's go Dutch.

Plumage: Bowling shirts with someone else's name; bell-bottoms, leisure suits and other fashions from the seventies that still have wear in them

Habitat: All-you-can-eat buffets; lounges at happy hour featuring two-fers and free hors d'oeuvres; discount matinees

Behavior Display: After a restaurant meal, nudges check toward center of table. Fumbles with wallet. Prone to frequent absent-mindedness in forgetting to bring any cash or leaving his wallet in his other plumage. Gifts presented to female during courtship include towels embroidered with hotel chain names, tiny bottles of shampoo and travel sewing kits, flower arrangements from banquets, promotional T-shirts from time-share condo resorts where this species takes free vacations.

Wallstreeter

Pin-Striped Yupster

Plumage similar in all three subspecies—three-piece suit (frequently Brooks Brothers), buttoned suspenders, silk tie, Coach briefcase, five-hundred-dollar wingtips or tasseled loafers. All three subspecies frequently nest in fern bars.

Subspecies: **Wallstreeter**

Latin Name: *Mastera universus*

Call: A curious dialect of paired words—Dow-Jones; Standard-Poor; bull-bear; buy-sell; bid-ask; put-call

Habitat: High-rise office buildings, trendy bars and restaurants in business districts of major metropolitan areas

Behavior Display: Usually performed with telephone held to ear. Frenzied activity keyed to stock quotes. Subject to mood swings related to fluctuations in the market. Occasionally talks about bizarre items like pork bellies. Though mostly obsessed with the price of stocks at this very moment, some are into futures.

Briefcased Lawbird

Pin-Striped Yupster

Subspecies: **Briefcased Lawbird**
(formerly Legal Eagle)

Latin Name: *Habeas corpus delecti*

Call: Sue! Sue!

Habitat: Courthouses and courtrooms; professional offices with multiple surnames on nameplate.

Behavior Display: This carrion eater has a cocky attitude and is highly argumentative. Constantly checking clock while talking to clients. Nonphysical threat display involves promises to see someone in court. Engages in frequent battles, but whether the judgment goes for or against his client, he always comes out the winner.

Common Businessman

Pin-Striped Yupster

Subspecies: Common Businessman

Latin Name: *Profita motiva*

Call: What's the bottom line?   35

Habitat: Businesses and offices throughout America

Behavior Display: Interested in P&Ls and balance sheets. Constantly complains about government interference. Votes Republican. Display involves flurry of activity, especially at the end of the month, to increase revenues, decrease expenditures or hide income. Mate often well-clothed and fed, but must settle for infrequent mating due to this species' constant exhaustion.

Ex-Tinct Husband

Ex-Tinct Husband

Latín Name: *Spousal exodus*

Call: The check's in the mail.

Plumage: Much less attractive than it used to be

Habitat: Found everywhere but becomes invisible at the hint of responsibility

Behavior Display: This species of albatross visits the nest two nights a week and every other weekend to interact with young. Asks prying questions about ex-mate's life, finances and new boyfriend. Some individuals are only stimulated to activity by court orders. Those who hang about the nest can be put to flight with the words "maintenance" and "child support."

Dodos

40

Tool-Belted Fixit

Tool-Belted Fixit

Latin Name: *George bungleitis*

Call: I can fix it, honey.

Plumage: Leather tool belt festooned with hammers, screwdrivers, levels. Red-scorched face (from lighting the pilot light), bandaged thumbs (from being hit by hammers), singed eyebrows (from soldering pipes).

41

Habitat: Homes where nothing works

Behavior Display: Eyes light up at the opportunity for home repair. Displays include flooding the house by failing to turn off the water before working on the plumbing; blowing out all the lights while trying to fix an electrical outlet; bashing holes in the drywall by missing nail with hammer; setting fire to the laundry while soldering a pipe; lowering the height of the dining table to six inches while trying to even the legs.

Cowbobolink

Cowbobolink

Latin Name: *Achy breakus*

Call: Howdy, there, little filly!

Plumage: Wide-brimmed hat, blue jeans, high-heeled boots, large belt buckle and tooled leather belt bearing individual's name. Field mark is worn circle on hip pocket of jeans denoting presence of a can of Skoal. Similar species found in urban area country-western bars mimics the plumage, but boots are shiny and new. Genuine article identifiable by traces of cow manure on boots.

Habitat: Feed stores, pickup trucks, rodeos and western bars

Behavior Display: Nighttime display involves performing two-step ritualized dance to twangy country-western music. During the day, drives pickup with dog riding in back. Preferred music relates trite, overly simplistic she-done-him-wrong tales. Very polite to females but refers to them as livestock—filly, heifer, lamb. Country-boy charm effective in winning female during courtship; later attitude toward her similar to that shown a recalcitrant steer.

Trustfund Booby

Trustfund Booby

Latin Name: *Silver spooner*

Call: Father, I'm a little short of funds ... .

Plumage: Anything from the L.L. Bean catalog

Habitat: Trendy, expensive nightclubs and restaurants; marginal small business ventures bankrolled by daddy

Behavior Display: Flits about in endless cycle of meaningless social engagements. Charges a great deal of merchandise but never pays its bills. Usually bears a single-syllable name like Kit, Chip, Kip or Biff. Prefers to mate only with females of its own social class. Though this species has physically left the nest it is still being brooded by parents financially.

Hard-Hatted Hammerer

Hard-Hatted Hammerer

Latin Name: *Constructus erector*

Call: A whistling "wheet whoo"

Plumage: Rolled up sleeves to expose arm muscles and protective hard hat from which the species derives its name. Leather tool belt slung low around hips causes pants to ride down, exposing tail cleavage.

47

Habitat: Always found moving busily about partially constructed buildings, hammering, ascending ladders or directing activities of large equipment. Groups of *C. erector* frequently gather around one spot, several of them seeming to watch idly as one individual works.

Behavior Display: Display, usually stimulated by passage of a female, regardless of her age, dress or attractiveness, includes ogling stare, temporary halting of activity, pushing back of hat, accompanied by vocalization of the familiar "wheet whoo" call.

48

Redneck

Redneck

Latin Name: *Gunrackus inbackus*

Call: My woman, yes; my dog, maybe; my pickup, never!

Plumage: Jeans or overalls, boots, beer belly, ball cap

Habitat: Pickup truck with gun racks and NRA decals. Rural areas of America.

49

Behavior Display: Life habits of this species can be understood by listening to country-western music. Flocks with persons named Joe Bob at local diner for morning coffee. Reappears at local honky tonk in evening to imbibe Lone Star longnecks. Mate is often also his cousin. Female roosts next to male on bench seat of pickup.

Not-Too Swift

Not-Too Swift

Latin Name: *Nobrainer trogladytus*

Call: Wo-man

Plumage: T-shirt worn inside out, mismatched socks, trouser fly often down. Behavioral field mark is inability to walk and chew gum at same time.

51

Habitat: Comic book rack at grocery store

Behavior Display: Knuckles drag on ground. Puzzled look on face. Unable to comprehend words of more than two syllables; must be spoken to slowly. Uncomfortable in the presence of females who can form a complete sentence. Curiously, for years this species has held positions of authority and been promoted over women who qualify for Mensa.

Workaholic Warbler

Workaholic Warbler

Latin Name: *Getta lifeus*

Call: I won't be home for dinner, honey. I have to work.

Plumage: Rolled-up sleeves, silk tie, corporate look. Eyes squint from constantly peering at computer screens and balance sheets.

Habitat: The Office—days, evenings, nights, weekends, holidays

Behavior Display: Abandons mate and young for work setting. Nestlings don't recognize him. Unable to enjoy ordinary activities. Happy only when working. When on rare vacation, travels with laptop and portable fax and calls office daily. Typically absent from activities important to family (school plays, birthdays, wife's work functions, Christmas) because must be at office. Even misses marriage counseling sessions because he has to work. Fails to notice absence of mate until served with divorce papers.

Game Birds

Bonded Male

Bonded Male

Latin Name: *Itsa guything*

Call: Yo, Bubba, how's it hangin'?

Plumage: Team jerseys, beer-stained T-shirts

Habitat: Sports bars, athletic competitions

Behavior Display: Always appears in flocks, often in team uniform. Much more comfortable in company of other males than with females. Bonding occurs during course of some physical activity such as sports, hunting or drinking beer. Reinforced by arm-wrestling, punching, questioning sexual orientation, headbutting and telling of sophomoric jokes involving body parts and bodily functions. Rules of bonding forbid meaningful conversation. Display ends when wife calls or keg runs dry.

58

Surf Shooter

Surf Shooter

Latin Name: *Oceanus moondoggie*

Call: Yaaaa ... wipeout! (followed by a huge splash)

Plumage: Shaggy blonde hair; white nose; bare chest, legs and arms. Summer plumage—baggy, brightly colored jam-type shorts; winter plumage—black or brightly colored wetsuit. Surfboard appendage tethered to one ankle.

Habitat: This marine species frequents sandy shores and oceanfront beaches of coastal North America, wherever the surf break is six inches or higher.

Behavior Display: Standard display, repeated to point of exhaustion, involves paddling out to sea in a prone position on top of a surfboard, bobbing on incoming waves, then suddenly jumping to the feet for a brief, erratic ride before making major wipeout into breaking surf. Highly territorial; flocks of *O. moondoggie* stake out beach territories where they spend periods of inactivity roosting with beer, burgers, babes and rock and roll music. Hierarchy based on board length. Speaks strange dialect of phrases involving "shooting the curl," getting "tubed," "hanging ten." This species worships its own god, the Big Kahuna.

Bulge-Breasted Bodybuilder

Bulge-Breasted Bodybuilder

Latin Name: *Pectoralis deltoides*

Call: A grunting "unh unh"

Plumage: A scrap of fabric draped over the torso, artfully ripped to expose arms, shoulders and chest muscles. Breast usually unfeathered. Broad leather belt band around middle. Spandex shorts on lower body. Feather coloration black alternating with neon orange, yellow or green.

61

Habitat: Found in health clubs and gymnasiums, roosting among the barbells and weight equipment

Behavior Display: Exhibited in presence of females or other competing males, the common display is a flexing of the upper body, including expansion of chest and flexion of arm and shoulder muscles. Reflection in mirror also elicits flexing display.

62

Ball-Bearing Jocko

Ball-Bearing Jocko

Latin Name: *Cleatsand sweatsox*

Call: Let's kick sum butt!

Plumage: Numerous plumage variations, depending on activity, but all have number patterns on back and wings. Some subspecies have hard, round heads and padded shoulders; most have cleated talons; all have jockstraps. Always carries a ball of some shape or size.

Habitat: Baseball diamonds, football fields, gymnasiums or any habitat where sports are played. Nest is strewn with a collection of bats, balls, pads, sticks and athletic shoes.

Behavior Display: Flocks with other jockos to engage in competitive, territorial contests. In this species, dominance interactions are highly ritualized, occurring in a specific setting with a series of rules governed by designated officials. Contests include hitting at spherical objects with sticks and pursuing round or ovoid balls down a field, court or icy surface. These tests of athletic prowess serve as displays of manhood and breeding fitness. Secondary displays such as spitting, crotch-scratching, ball-spiking and patting tail feathers of team members are also very important. Scoring is highly valued, both on and off the playing field.

Couched Potato

Couched Potato

Latin Name: *Sofasprawl spudensis*

Call: I'll do it later, honey.

Plumage: Undershirt, boxer shorts and short black socks

Habitat: Living rooms across America

Behavior Display: Highly sedentary species adopts a supine posture on living room furniture. Incessant television viewing accompanied by feeding frenzy and channel grazing. Interrupts this behavior only for trips to the kitchen and bathroom. Elicits exasperated response from mate due to chores constantly left undone.

Snowbird

Snowbird

Latín Name: *Mogul masherii*

Call: Single! Single!

Plumage: Brightly colored plumage changes yearly. Field mark is raccoonlike facemask in shape of ski goggles. Feet have evolved into long, flat, boardlike appendages. Many have surgical scars on knees.

67

Habitat: Snowy mountain slopes, chairlifts and après-ski bars

Behavior Display: Active in winter. Obsessively listens to weather and reads ski reports, seeking fresh powder in the back bowls. Display triggered by double black diamond trail sign, involves skiing downhill as fast as possible in attempt to break neck. This species becomes predatory at night, hunting snow bunnies. Any mating occurring at ski resort is soon forgotten.

Cheering Sportsfinch

Cheering Sportsfinch

Latin Name: *Armchair quarterbackus*

Call: How 'bout them Broncos!

Plumage: Athletic jersey or T-shirt and ball cap emblazoned with the name of a professional sports franchise. Plumage usually bears the name of an aggressive animal species such as Lion, Tiger or Bear; this may be in the diminutive form such as Cub.

69

Habitat: Any sporting event, from football game to tractor pull

Behavior Display: Watches any and all sporting events, including curling, archery, Australian rules football, and monster trucks. Has trouble spelling ESPN, but knows all the sports scores and the history and stats of all professional players. Clutching can of beer an essential part of spectator display. Always knows how to play the sport better than the players and doesn't hesitate to announce how it should have been done. Also knows rules and sees plays better than all officials. Ignores mate and nestlings if sporting event is on television. Particularly active (actually, inactive) on Thanksgiving and New Year's Day.

Fishorfowl Finch

Fishorfowl Finch

Latin Name: *Hookand bulletus*

Call: Mimics calls of waterfowl, elk and other game species

Plumage: Bright orange or camouflage plumage, depending on the season. Head plumage of *hook* subspecies adorned with small, hooked "flies." Appendage always attached to rod or rifle.

71

Habitat: Forests, fields, stream sides. When in the field, always returns to location of cases of beer.

Behavior Display: Highly predatory. During fall season, stalks through wooded areas and fields, shooting at anything that moves (even others of its own species). In summer, stands hip-deep in rushing stream waving fishing pole in the air or sleeps on lakeshore with fishing pole suspended over water. Bragging display involves flocking with others of species, imbibing malt beverages and fabricating stories of prowess involving "ones that got away."

Teebird

Teebird

Latin Name: *Bogey birdie*

Call: Fore!

Plumage: Brightly colored plumage in unexpected combination of brassy yellow, grass green and candy apple red, topped with plaid cloth hat. Alligator marking on left breast. Scottish subspecies wears baggy knickers and Argyle socks. Carries large bag over shoulder containing sticks made of wood and iron.

73

Habitat: Golf courses, discount golf shops, 19th Hole Bar

Behavior Display: Obsessively hits small white balls around simulated natural habitat. More interested in holes-in-one on the course than on the homefront. Abandons mate and nest each weekend during daylight hours to flock in groups of two to four. Phobic fear of sand and water. Temper tantrums follow "going over par." Display ends with detailed recitation of the round just played. This species speaks in its own strange language, using terms like double bogey, dogleg and back nine. Often chirps about his handicap.

Carnary

Carnary

Latin Name: *Fuelus injected*

Call: Great wheels, babe!

Plumage: Oil stains on feathers. Permanent grease under talons. Tail plumage resembles oily rag.

Habitat: Under the hood of a car

75

Behavior Display: Obsessively tinkering with cars. Mutters about RPMs, MPGs, 396s, zero to 60s. Less interested in female than in car she drives. Prone to wander off during date to check out a Ferrari Dino or a '65 GTO. Likes his females with twin cams. Mesmerized by auto exhaust and sound of a revving engine.

Songbirds

Sensitive New-Age Sparrow

Sensitive New-Age Sparrow

Latin Name: *Alan aldas*

Call: I want to share my feelings with you.

Plumage: Water-resistant to repel tears

Habitat: Counseling groups, poetry readings, ballet classes

Behavior Display: Helps with nest building and rearing of hatchlings. Not only brings worms and grasshoppers home to the nest to feed young, but helps do the dishes afterwards. Talks with mate about feelings. Understands the meaning of the word "share" without consulting a dictionary. Belching, flexing and bragging noticeably absent from this species' display. Eats quiche.

80

Blushing Shyguy

Blushing Shyguy

Latin Name: *Womanus avoidii*

Call: Unintelligible stutter

Plumage: Unremarkable plumage designed to blend into the background and make this species as unnoticeable as possible. Reddened face and neck sure sign of presence of female.

Habitat: Found in multiple habitats, yet difficult to detect due to secretive nature

Behavior Display: Blushes and faints at female's approach. Once spotted, utters call and disappears. Bashful nature of shyguy elicits fascination and pursuit behavior in female. However, many of this species have never mated.

Spectacled Scholar

Spectacled Scholar

Latin Name: *Professorus studyi*

Call: Usually silent, but when interrupted utters a baffled, distracted "Huh?"

Plumage: Tweed plumage, generally rumpled; traditional English country look. Wire or horn-rimmed glasses common. Many are bearded.

Habitat: Libraries, dens, offices. Nest is always cluttered with papers, journals and textbooks.

Behavior Display: This species spends much of the day with its beak buried in a book. The presence of fledglings of any species stimulates *P. studyi* to give lectures on esoteric topics. Tends to display apathy or bewilderment in the presence of females, resulting in heightened female interest. Elaborate display with filling, lighting and smoking a pipe common.

Computerated Hummingnerd

Computerated Hummingnerd

Latin Name: *Ramrom megabyteus*

Call: Boot up!

Plumage: Coloration very pale from being indoors in the glow of a computer screen. Field mark is plastic pocket protector.

Habitat: Virtual reality

85

Behavior Display: This species is always found perched in front of computer staring at screen. Continuously flitting about in search of bits and bytes. Nest is always littered with floppy disks and copies of *PC* magazine. Flock members communicate via modem. Infantile behavior is characteristic of this species, in sharp contrast to its intellectual acuity. Prone to hover in flocks at computers, devising juvenile names for computer commands and accessories. Recreation consists of playing Nintendo. This species does not have breeding success until it becomes a multimillionaire by starting its own computer company, turning its floppy disk into a hard drive.

Nice Gull

Nice Gull

Latin Name: *Yourmotha lovesim*

Call: You look very nice tonight.

Plumage: Matched socks, neatly ironed shirt, lack of food stains on clothes

Habitat: Exactly where he says he'll be, and on time. This species is often good at nest-building but can't seem to attract a mate.

Behavior Display: Often found in company of females but mating rarely occurs. Frequently listens to female's problems with other males. This species treats females courteously and with respect, showers before a date, is never late, calls when he says he will, listens attentively, does his own laundry and doesn't hog the channel selector. Curiously, this results in loss of interest by female. Attempts at courtship result in typical response, "I really like you as a friend." However, mothers always find this species sweet and adorable.

Bird of Paradise

Bird of Paradise

Latin Name: *Gentlemen's quarterlarius*

Call: Shaken, not stirred.

Plumage: Plumage of male much more flamboyant than the female. Exceptionally well-tailored, European-cut suits. Field mark is designer label plumage pattern.

Habitat: Abercrombie & Fitch

Behavior Display: Constantly preening. Assumes dramatic poses, usually to display strong jawline. Constantly peers at own reflection in plate glass windows and the mirrored sunglasses of those with whom he is conversing. Routinely spends more time in preparation for an outing than his date does. Utters foreign terms—Armani, Yves St. Laurent—for effect. Favorite pet, a Jaguar.

Tribal Turkey

Tribal Turkey

Latin Name: *Malus reclaimus*

Call: A shouted "Ho!" followed by drumbeats

Plumage: Loincloth, bare chest painted with pseudo-tribal designs (in tempera paints from Walgreen's), feathers stuck in hair. Primitive costume is completed by black shoes and socks or high-top Nikes.

Habitat: Men's movement seminar

Behavior Display: Flocks sit cross-legged in circle and beat on drums. Fire-starting and chanting incoherent grunts very important. Frequent crying and hugging and attempts to get in touch with the inner fledgling. Much discussion of how all life's problems stem from their fathers' failure to play baseball with them.

Universal Species

Mythical Dreambird

Mythical Dreambird

Latin Name: *Masculoid perfectus*

Call: Be mine.

Plumage: This species has incredibly good-looking plumage but doesn't know it. Plumage varies from tall, dark and handsome to blond, blue-eyed and boyish.

95

Habitat: In your dreams

Behavior Display: This species is incredibly sweet and nice, while also sexy, fascinating and oozing chemistry. His display includes accompanying his mate to parties, dancing and shopping and planning fun, romantic outings. The dreambird helps around the house and is always warm, caring and sensitive. He is open, honest and talks about his feelings. At the same time he is manly, muscular and good at all sports. Excellent at interpreting nonverbal signals of the female. Encourages mate in her career and brags about her to other males. Earns six-figure income and loves dogs and children. Mysterious similarity to Donna Reed's husband.

Typical Male

Typical Male

Latin Name: *Y-chromosomal americanus*

Call: You're not mad, are you?

Plumage: May appear in jeans, T-shirt and white high tops or suit and tie.

Habitat: Our homes, our lives, everywhere we turn

97

Behavior Display: Never asks for directions. Only interested when the female is hard to get. Doesn't understand meaning of "I'll call soon." Incapable of putting down the toilet seat. Asks female out for date one hour before and expects her to be available. Typical response to female's request to talk about the relationship is "Just let me see the end of the game." Good at winning female over with boyish charm after screwing up.

Male-Watcher's Species Checklist

Species	Time/Date of Sighting	Behaviors Observed	Did You Get Hit On?
Birds of Prey			
Noncommittal Nuthatch	_____	_____	_____
Heron Chest	_____	_____	_____
Valentino's Warbler	_____	_____	_____
Bikerhawk	_____	_____	_____
Midlife Chrysler	_____	_____	_____
Horned Stud	_____	_____	_____
Greater Spender	_____	_____	_____
Disco Kinglet	_____	_____	_____
Woodypecker	_____	_____	_____
Bluejerk	_____	_____	_____
Male Chauvinist Pigeon	_____	_____	_____
Married Magpie	_____	_____	_____
Carrion Eaters			
Black-Capped Cheapadate	_____	_____	_____
Wallstreeter	_____	_____	_____
Briefcased Lawbird	_____	_____	_____

Species	Time/Date of Sighting	Behaviors Observed	Did You Get Hit On?
Common Businessman	_____	_____	_____
Ex-Tinct Husband	_____	_____	_____

Dodos

Tool-Belted Fixit	_____	_____	_____
Cowbobolink	_____	_____	_____
Trustfund Booby	_____	_____	_____
Hard-Hatted Hammerer	_____	_____	_____
Redneck	_____	_____	_____
Not-Too Swift	_____	_____	_____
Workaholic Warbler	_____	_____	_____

Game Birds

Bonded Male	_____	_____	_____
Surf Shooter	_____	_____	_____
Bulge-Breasted Bodybuilder	_____	_____	_____
Ball-Bearing Jocko	_____	_____	_____
Couched Potato	_____	_____	_____
Snowbird	_____	_____	_____
Cheering Sportsfinch	_____	_____	_____
Fishorfowl Finch	_____	_____	_____
Teebird	_____	_____	_____
Carnary	_____	_____	_____

Species	Time/Date of Sighting	Behaviors Observed	Did You Get Hit On?
Songbirds			
Sensitive New-Age Sparrow	_____	_____	_____
Blushing Shyguy	_____	_____	_____
Computerated Hummingnerd	_____	_____	_____
Nice Gull	_____	_____	_____
Bird of Paradise	_____	_____	_____
Tribal Turkey	_____	_____	_____
Universal Species			
Mythical Dreambird	_____	_____	_____
Typical Male	_____	_____	_____

Mary Taylor Gray

is an avid watcher of both birds and males, and she finds both fascinating, though birds are much easier to understand. She is the author of *Watchable Birds of the Rocky Mountains* (the real thing) and writes a monthly column on birdwatching for the *Rocky Mountain News*. She lives in Denver.